THE LIBRARY OF THE PLANETS™

VENUS

Amy Margaret

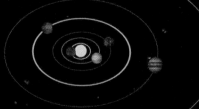

The Rosen Publishing Group's
PowerKids Press™
New York

For Marisa, Evan, Emily, and little Alex

Published in 2001 by The Rosen Publishing Group, Inc.
29 East 21st Street, New York, NY 10010

Copyright © 2001 by the Rosen Publishing Group, Inc.

First Edition

Book Design: Michael Caroleo and Michael de Guzman

Photo Credits: Cover, title page, and pp. 4, 7 (Venus), 8 (Venus), 11 (space backdrop), 15, 17 (Venus) PhotoDisc; pp. 7, 11 (Illustrations) by Mike Caroleo; p. 7 (Galileo) © Archive Photos; pp. 8 (space probes), 19 Courtesy of NASA/JPL/California Institute of Technology; p. 12 (Illustration) by Maria Melendez; p. 16 (greenhouse) © Joseph Sohm; ChromoSohm Inc./CORBIS; p. 20 Michael Whelan/NGS Image Collection; p. 20 (view from Hawaii) © Gary Braasch/CORBIS.

Margaret, Amy
    Venus/by Amy Margaret
      p. cm.—(The library of the planets)
    Includes index.
    ISBN 0-8239-5643-1 (lib. bdg. : alk. paper)
      1. Venus (Planet) I. Title. II. Series.
    QB621 .M36 2001
    523.42—dc21
                                                    99-053833

Manufactured in the United States of America

Contents

Venus, the Universe's Hot Spot

For hundreds of years, Venus has been called the sister planet or twin planet of Earth. This is because Venus is about the same size and weight as planet Earth. Through space exploration in the last 30 years, scientists have found that the two planets are actually very different from each other.

Venus is the second planet from the Sun. Venus is also the hottest planet in our **solar system**. Our solar system is made up of the Sun, the nine planets, and their moons. Our solar system is really only a small part of a larger system. This system is called the Milky Way **galaxy**. A galaxy is a large group of stars. Any of these stars may have planets revolving around them. Many galaxies make up the entire **universe**. The planets in our solar system may have been formed through an explosion called the Big Bang. The Big Bang probably occurred about 10 billion years ago. It took several billion years after that for our solar system to be created.

This image of Venus is made up of pictures taken by a space probe. A space probe is a spacecraft that travels in space and is steered by scientists on the ground. Behind Venus is an image of the Milky Way.

The History of Venus

Venus was first seen in ancient times. It was named after the Roman goddess of love and beauty. This is because the planet shined so brightly in the sky. **Astronomer** Galileo Galilei studied Venus in 1610. Galileo used one of the earliest **telescopes**. With the telescope, Galileo discovered that Venus has phases that make its appearance seem to change. When Venus is between the Sun and Earth, it cannot be seen. The Sun lights different parts of Venus as the planet **orbits** the Sun. Venus first looks like a thin banana shape, called a crescent. It then looks like a half circle. Finally, it looks like a full circle. Once Venus becomes a full circle, the Sun is in between Earth and Venus. Again, Venus will not be seen from Earth. As it travels back around the other side of the Sun, Venus looks like it is getting thinner again. From Earth, Venus is easiest to spot when it is in its crescent shape. This is because Venus is closest to our planet at that time.

Galileo Galilei studied Venus in the early 1600s. The picture at the bottom of the page shows the different phases of Venus as it orbits the Sun.

Exploring Venus From Space

In 1961, Russia tried to send the first **space probe** to Venus. A space probe travels outside of Earth's **atmosphere** to study the solar system. It does not carry people. The Russian probe lost contact with the scientists on the ground. Nobody knows what happened to it. In 1962, the United States **launched** *Mariner 2,* a space probe that flew past Venus. *Mariner 2* measured Venus's surface **temperature**. It was almost 900 degrees Fahrenheit (482 degrees C). Only a few space probes have actually landed on Venus. In less than an hour, the heat and atmosphere destroyed them! In 1978, the United States space probe *Pioneer Venus* used computers to make maps of the planet's surface. These were the most recent pictures of Venus we had until the *Magellan* mission in 1990.

The image on the left is of the Mariner 2 space probe as it might look circling Venus. On the right is a computer-made image of the Magellan space probe going around the planet.

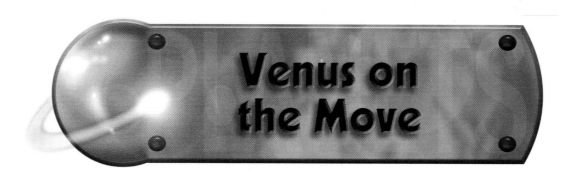

Venus on the Move

Every planet spins on its **axis** similar to the way a toy top spins. Venus **rotates** differently from most planets, turning clockwise, east to west. Most of the other planets spin west to east. Venus moves very slowly on its axis. It makes a complete rotation in 243 days. Earth takes only 24 hours, or one Earth day. The Sun pulls the nine planets toward itself, making the planets circle the Sun. The farther away a planet is from the Sun, the longer it takes to orbit it. Venus also moves around the Sun. It circles the Sun in 225 days. This equals a year on Venus. That means a year on Venus is shorter than a day on Venus!

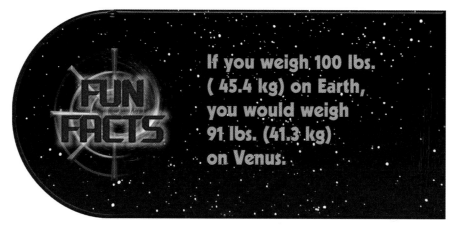

FUN FACTS

If you weigh 100 lbs. (45.4 kg) on Earth, you would weigh 91 lbs. (41.3 kg) on Venus.

This is a drawing of Venus circling the Sun.

FUN FACTS

Planet	Orbit Time Around the Sun
Mercury	88 Earth days
Venus	225 Earth days
Earth	1 year (365 days)
Mars	1 year and 322 Earth days
Jupiter	12 Earth years
Saturn	29 1/2 Earth years
Uranus	84 Earth years
Neptune	165 Earth years
Pluto	248 1/2 Earth years

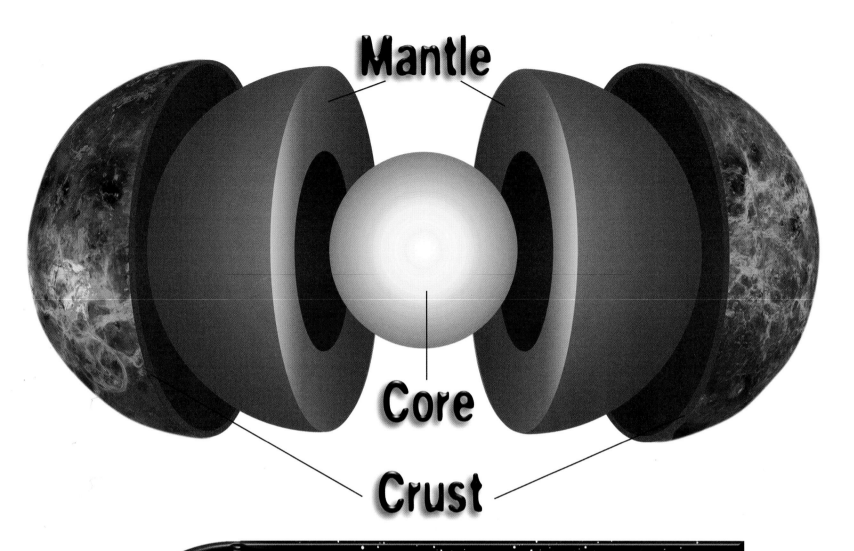

Mantle

Core

Crust

The Magnetic Mystery
Venus is believed to spin too slowly on its axis to create a magnetic field. Mercury, the planet closest to the Sun, is magnetic, but its rotation is almost as slow as Venus's. Scientists do not know what makes Mercury magnetic and Venus not magnetic.

Inside Venus and the Inner Planets

Scientists divide the nine planets into two groups called the inner planets and the outer planets. Venus, Mercury, Earth, and Mars are called the inner planets because they are closest to the Sun. Each of these planets has a hard surface. This surface is called a crust. The inner planets also have a middle section called the mantle. The center of each inner planet is called the core.

The cores of Venus, Mercury, and Earth are made of iron. As Earth and Mercury spin, the iron moves. This creates a magnetic field. You can see magnetism at work when you put two refrigerator magnets together. They will either push away from or pull toward each other. Venus is the only inner planet with little or no magnetism. This may be because the planet spins so slowly. Scientists also think Venus's iron core may be too small to create any magnetism.

The hard outer surface of Venus is called the crust. The middle layer is called the mantle, and the center, made of iron, is known as the core.

Comparing the Sister Planets

Earth is 7,930 miles (12,762 km) in **diameter**, and Venus is 7,520 miles (12,102 km) in diameter. Both planets have volcanoes, valleys, and canyons on their surfaces. Earth and Venus have very few **craters**. In most other ways, though, the two planets are very different. Venus's atmosphere is made up mostly of carbon dioxide. This gas is poisonous if we breathe it. Lightning storms beat down on Venus all the time. The atmosphere contains many layers of clouds that completely hide Venus's surface. These clouds keep the temperature on Venus extremely hot, making it the hottest planet in our solar system.

The gravity on Earth keeps us from floating off into space without pushing us into the ground. Air pressure is the amount of force that the atmosphere puts on a planet. On Venus the pressure is strong enough to crush people, cars, and even buildings.

These are computer images of Earth (left) and Venus (right). Below each planet is an actual picture of the mountainous surface of Earth and Venus. These two planets are very close in size. They are not this close together in space, though.

WHERE'S THE WATER?
At one time, Venus may have had large bodies of water, like Earth. If so, the Sun boiled the water away. Now Venus is a dry planet. If Earth had been closer to the Sun, it might have had the same fate!

The Greenhouse Effect on Venus

The **greenhouse effect** takes place when heat from sunlight is trapped under the glass of a greenhouse. A greenhouse is a warm place where plants are grown. It is made of glass so sunshine can reach the plants. The sunshine raises the temperature inside the greenhouse. This creates healthy conditions for plants to grow. The greenhouse effect also takes place on Venus. Thick clouds that cover the planet act like the sheets of glass. Some of the Sun's rays go through the clouds to the planet's surface. The rays bounce back up into the atmosphere. They cannot pass through the heavy cloud layer. This traps the heat, raising the temperature on Venus's surface. The blazing temperature on Venus keeps water from forming there today. Without water, there can be no life. The **intense** heat also keeps humans from ever being able to land on this planet.

The greenhouse effect causes temperatures to rise very high on Venus.

The Dangerous Surface of Venus

Scientists used **radar** pictures taken by the *Magellan* space probe in 1990 to piece together an almost complete picture of Venus's surface. Radar uses radio waves that can go through the thick clouds on Venus to find land features. Venus appears to have a mostly flat surface. However, Venus does have two areas that contain mountains and canyons. The Ishtar Terra is the bigger of the two areas. The Ishtar Terra is a little larger than the land area of the United States. The second area, the Aphrodite Terra, is about half the size of the continent of Africa.

Radar images from *Magellan* show tens of thousands of volcanoes on Venus's surface. **Lava** can spray from these volcanoes. Volcanic rock covers much of Venus. Most of the volcanic activity probably happened about 500 million years ago. Scientists have recently discovered that Venus still has a few active volcanoes.

This map shows the surface of Venus. The image was taken by the Magellan *space probe.*

Earth's Moon

Venus

CHECK IT OUT!
To find out where Venus is in its orbit around the Sun, look on the Internet. This address will help you pinpoint its location: http://www.skypub.com/sights/sights.shtml

FUN FACTS

Seeing Venus From Earth

Venus is the easiest of all the planets to spot in the sky without a telescope. It is brighter than most of the stars in the sky. The Sun and the Earth's Moon are the only objects in space that are easier to see than Venus.

When Venus is to the left of the Sun, you can see it in the morning, just before the Sun rises. When Venus is to the right of the Sun, it can be seen in the evening, after the Sun sets. This is why Venus has often been called the morning star or the evening star.

With a strong telescope you can see the thick clouds around Venus. These clouds give Venus its bright appearance, because they reflect a lot of light from the Sun. Remember that when looking for Venus or other objects in the sky, you should never look directly at the Sun. It could cause permanent eye damage or even blindness.

This picture of Venus was taken from the ground in the state of Hawaii. On the right is a picture of the goddess Venus. The Romans named the planet Venus after this model of love and beauty.

The Magellan Mission and Beyond

We know much more about Venus than ever before because of NASA's *Magellan* mission. *Magellan* was sent to orbit Venus on May 4, 1989. It arrived in Venus's atmosphere in the summer of 1990.

Cameras were not used on this mission because the planet's cloud cover is impossible to see through. It took *Magellan* about eight months to pick up radar **scans** of most of Venus. Its equipment was new, so the images are clearer than those taken on earlier missions.

It is not possible for humans to walk on Venus to study it up close. We will continue to learn more about Venus through future space probes. These probes will explore this hot, dangerous planet from a safe distance.

Glossary

astronomer (ah-STRAH-nuh-mer) One who studies the night sky, the planets, moons, stars, and other objects found there.

atmosphere (AT-muh-sfeer) The layer of gases that surrounds an object in space. On Earth, this layer is the air.

axis (AK-sis) A straight line on which an object turns or seems to turn.

craters (KRAY-terz) Holes in the ground that are shaped like bowls.

diameter (dy-A-meh-tehr) The length through the center of an object.

galaxy (GAH-lik-see) A large group of stars and the planets that circle them.

greenhouse effect (GREEN-hows eh-FEKT) A scientific process that occurs when heat from sunlight is trapped under the glass of a greenhouse. This can also take place on a planet.

intense (in-TENTS) Very strong.

launched (LAWNCHED) Pushed out or put into the air.

lava (LAH-vuh) A hot liquid made of melted rock that comes out of a volcano.

orbits (OR-bits) When a planet circles around another object.

radar (RAY-dar) A system that uses radio waves to find objects or land features.

rotates (ROH-tayts) When something moves in a circle.

scans (SKANS) Images produced by a computer.

solar system (SOH-ler SIS-tem) A group of planets that circle a star. Our solar system has nine planets, which circle the Sun.

space probe (SPAYS PROHB) A spacecraft that travels in space and is steered by scientists on the ground.

telescopes (TEL-uh-skohps) Instruments used to make distant objects appear closer and larger.

temperature (TEM-pruh-cher) How hot or cold something is.

universe (YOO-nih-vers) Everything that is around us.

Index

A
atmosphere, 9, 14, 17, 22
axis, 10

C
core, 13
craters, 14
crust, 13

G
galaxy, 5
Galilei, Galileo, 6

M
mantle, 13
maps, 9
moons, 5, 21

O
orbits, 6, 10, 22

R
radar, 18, 22

S
scientists, 5, 9, 13, 18
solar system, 5, 9, 14

T
telescopes, 6, 21
temperature, 9, 14, 17

U
universe, 5

V
volcanoes, 14, 18

If you would like to learn more about Venus, check out these Web sites:
http://www.tcsn.net/afiner
http://www.personal.vineyard.net/jamie/msnctrl.htm